PUSHKI

Although n̶ ̶e̶s̶ire to fly has existed for as long as we have (Daedalus and Icarus, the flying chariots of Sanskrit epics, Leonardo da Vinci's *Codex on the Flight of Birds*), the modern era of flight did not begin until 1783, with the demonstration of the first working hot-air balloon by the Montgolfier brothers. The first passengers were a duck, a sheep and a rooster.

Air travel went from strength to strength, and Nadar's sixty-metre high balloon, Le Géant, was the biggest balloon ever built. Requiring 300 seamstresses to assemble its 22,000 yards of silk, it was not a success. But it paved the way for thinking on heavier-than-air flight—which would eventually lead to the "the first sustained and controlled heavier-than-air powered flight" in 1903 by the Wright brothers.

The books in "Found on the Shelves" have been chosen to give a fascinating insight into the treasures that can be found while browsing in The London Library. Now celebrating its 175th anniversary, with over seventeen miles of shelving and more than a million books, The London Library has become an unrivalled archive of the modes, manners and thoughts of each generation which has helped to form it.

From essays on sherry and claret to a passionate defence of early air travel, from advice on how best to navigate the Victorian dinner party to accounts of European adventures, they are as readable and relevant today as they were more than a century ago—even if flying is nowadays seen by many not just as a human right, but as a human necessity.

THE RIGHT TO FLY

The London Library

Pushkin Press

Pushkin Press
71–75 Shelton Street,
London WC2H 9JQ

Nadar, *The Right To Fly*, translated from the French by James Spence
Harry, with a Preface by George Sand. London: Cassell, Petter, and
Galpin, 1866

"Nadar's Giant Balloon at Paris", by an unidentified illustrator,
published in *The Illustrated London News*, October 10, 1863

First published by Pushkin Press in 2016

9 8 7 6 5 4 3 2 1

ISBN 978 1 782273 17 2

Set in Goudy Modern by Tetragon, London

Printed by CPI Group (UK) Ltd, Croydon, CR0 4YY

www.pushkinpress.com

PREFACE TO 'THE RIGHT TO FLY'

BY GEORGE SAND, 1865

AMANTINE-LUCILE-AURORE DUDEVANT (1804–1876) was a gifted French writer who adopted the pseudonym by which she is remembered for her novel *Indiana*, which brought her immediate fame. She scandalised society with her many lovers, among whom were Prosper Mérimée, Alfred de Musset, and Frédéric Chopin. The best known photograph of her was taken by her friend Nadar.

Truth has two modes of existence, marked by two distinct phases; that in which it is merely demonstrated, and that in which it can be proved.

In the former, it reposes first of all upon the faith which is the instinct of the Good and the Beautiful, then upon reason, and finally upon intellectual certitude.

In the latter it rests upon the experience acquired by the accomplished fact.

Honour to those men whose initiative sets free the first hypothesis, the sovereign induction, from the chaos of dreams, from the thousand gropings of the imagination struggling with the unknown! When those great and generous minds have succeeded in well posing the question to be solved, they have already made a grand step: they have opened the way.

Afterwards come the men of application, not less useful, not less admirable, who, by clever and patient experiments, proceed from the hypothesis to the discovery. From that time, Genius becomes a material force; and the idea which was only a promise becomes the real benefit with which the human race enriches itself.

Thus it has been with all our conquests in the domains of Science and of Industry. Every light has its precursory dawn, and he who perceives the one may predict the other. But all do not see the glimmer of the first breaking of a truth, and it is at this earliest stage of undecided brightness that it is disputed, sometimes even repulsed with passion, so formidable is the apparition of those great stars of progress which upset customary notions, destroy to a certain extent the world of the past, and, on the day when their rays burst forth above the horizon, cause man to enter upon new conditions of existence.

Thus it was with steam, with electricity, and with all those surprising inventions which, within the last century, have so essentially modified public and private life in and around us.

In the little work which follows our brief reflections, and also in the "Mémoires du Géant" (a book so innocently dramatic), read the protestations, the persecutions even, called forth by every truth passing through the phases of research and demonstration. It is when it stands most in need of consideration, study, and encouragement, that sarcasm and an impatient desire to crush it, rise up about the dawning truth. "Prove thyself!" is cried out on all sides, "and we will believe in thee."

Truth replies: "Assist me to ripen, to manifest myself. Give me the means for becoming a fact, and for that, know me; deny me not. I am but an idea, a soul, so to speak, and yet you expect to touch me before you have permitted me to take a body! I exist nevertheless; I exist in a sphere as real for the eyes of the understanding as if I were already a palpable fact. Respect me, alas! for if you deny me, you deny yourselves. I am yours, since I bring you the future; and if you affirm that I shall never be, you say that you yourselves desire never to exist."

Among the adepts, the ardent vulgarisers, and the devoted servants of Truth in the stage of demonstration, Nadar, who is neither a *savant* nor a speculator, but, in my opinion, a great logician and a man of firm will, here lays before us his earnest and deliberate Word.

This Word, summed up in the "Right to Fly," possesses a veritable worth and a veritable force. Let it be meditated without prejudice, and every serious mind will acknowledge, that the question of the "Right to Fly" is one of those magnificent questions which cannot remain unsolved from the moment that they are well posed.

GEORGE SAND.

PARIS, Nov. 2, 1865.

THE RIGHT TO FLY

BY FELIX NADAR, 1866

GASPARD-FÉLIX TOURNACHON (1820–1910) was a photographer, journalist, writer, left-wing agitator, caricaturist and aeronaut who was imprisoned for debts and inspired Jules Verne's *Five Weeks in a Balloon*. In 1858, he became the first person to take an aerial photograph, and in 1864 he photographed the Paris sewers for the first time.

Nadar's giant balloon at Paris

Note of the Translator

[The translation of the "RIGHT TO FLY" appears as the simple unpretending transmission of M. Nadar's work from French into English. It has been verified by the Author, who preferred occasional crudity of diction to the more complicated adaptation of his ideas to a foreign language, which would in some cases have detracted from the originality which characterises this little volume. The indulgence of the critics is therefore claimed, should the "well of English undefiled" appear in some cases to have been rendered somewhat impure.—J. S. H.]

Dedication

I dedicate the English Edition of this little book to the Society recently formed in England for the Study of Aerial Navigation (having His Grace the Duke of Argyle for its President, and the honourable and persevering Mr. Glaisher for its Secretary-General), under the patronage of the powerful British Association for the Advancement of Science.

I am particularly pleased to have this opportunity of expressing my profound satisfaction at the creation of this new Society.

Engaged in the common pursuit of such a humanising task, there should be no feeling of petty rivalry or jealousy; a spirit of generous emulation should alone animate the breasts of all the men of good will and good heart united for the interest of the Great Cause.

NADAR,
Founding Member of the International Society
of Encouragement for the Study of Aviation
or Aerial Locomotion by means of Apparatus
heavier than the Air (Paris, July 30, 1863).

PARIS, March 15, 1866.

To the Passer-By

You were passing. My title, well doing its business as a title, struck your eye; and here you are, stopped in your course.—You open me promiscuously.

Should you fall upon any hard word of the *savants*, don't be afraid, and stand firm!

There is no science here, since it is I who speak to you.

I ask you only for a grain of good sense, if you can supply it; ten minutes' attention, if you are capable of giving it; moreover, — and listen to this serious reason which generally touches you,—YOUR INTEREST is at stake in the matter.

Should you be somewhat of that opinion, when you have read me through,—if I have made you see what you did not dream of looking at,—if I have made you think of what you did not even suspect, then try to aid THE CAUSE—your own Cause! — and insult me for having been of service to you.

If you do not SEE, if you do not KNOW, if you do not BELIEVE, then never fly,—and continue to walk, simpleton!

I.

My estimable colleague, M. G. de la Landelle, returned one day from a visit to a most intelligent master clockmaker, who, in the wake of many predecessors, and, like many others, at a great loss of time and money had been hunting after a system of directive aerostation.

He had told him, time after time, and in every possible manner, that the "direction of balloons," properly speaking, is a chimera;—

—that upon those immense non-resisting surfaces, upon those enormous masses lighter than the volume of air they displace, the slightest current exercises a too easy influence;—

—that supposing even it were possible to apply to any description of balloon—round, conical, cylindrical, pisciform, annular, conjoined, or of any other shape—an apparatus of sufficient power (such as a screw-propeller, spirals, sails, and counter-sails, oars, paddles, rudders, &c.), the envelope of any such balloon, even if it were of steel or of triple brass, would be crushed by the first encounter—as an insect is crushed upon a nail—between two pressures, that of the current of air and that of the opposing apparatus;—

—that since the year of grace and deception 1783, since the *sublime and detestable* discovery of the Brothers Montgolfier, the endless series of discomfitures experienced by all the pretending "balloon directors" demonstrate historically that a balloon in the air is never and can never be anything else than a boat: that it is born a buoy, and will burst a buoy.

M. de la Landelle had tired himself out with repeating that neither birds nor insects, neither flying-squirrels nor bats, neither flying-lizards nor flying-fish, are Montgolfiers; and that they do not rise in the air by the effect of the hot air contained in their bodies—

—(As used to be very seriously affirmed by the Academy, now prudently silent thereanent, since it is incapable of acting.)

Without mentioning the interior heat of the

body of such insects as the Libellule and the Hydrophile, he had in fact proved to the worthy clockmaker, that if a bird rose in the air on the Montgolfierian principle, that is to say, by virtue of the "hot air" (according to the Academical doctrine) in its bones, feathers, &c., it would be absolutely necessary for the winged creature, in order to turn the provision of "hot air" to account in the ascension, to commence by inflating its envelope or volume seven or eight hundred-fold.

He had asked him why, if the bird rose through a simple difference of specific weight, the slightest shot in the extremity of its wing should suffice to cause it to sway and fall.

He had also asked him whether a bird weighs or not in its flight, when, in being let loose in a room, it comes into contact with and breaks the windows.*

With the most apostolical perseverance, he had proved to him superabundantly and *a fortiori*, that this pretended "hot air," far from assisting the bird

---

* J. A. Sanson, "Du Point d'Appui," 1841.

in its flight, would, on the contrary, *absolutely* prevent it;

—that, in every order of the animal scale, all the flying creatures rise, sustain themselves, and direct their course, not through the effect of a difference of specific weight, but through the admirable and eternal combination of the mechanical, statical, and dynamic laws;

—that a rocket also rises, in spite of its weight and of that of the motive force it carries with it;

—and that, finally, an object must be HEAVIER *than the air* (or more *dense*) to exercise an action upon the air, to command the air—in the same way as, in every order of things, it is indispensable *to be the strongest in order not to be beaten.*

All my friend's preaching produced no effect, alas!—it is so difficult for a new idea, however rational and evident it may be, to take the place of an old prejudice.

"A new truth," says Fontenelle, "is a wedge that only penetrates with time and by the wide end."

The paradox of yesterday becomes, however, the truth of to-morrow, and though the majorities of the morrow are always composed of the minorities of the day before, we know, to our grief, in all such cases, how much time these minorities require to become of age.

—And my friend La Landelle started homewards downcast and sad, and asking himself how many times it was necessary to repeat the same things in order to succeed in getting them to be understood.

All of a sudden, he encountered a ladder, which caused him to quit the pavement.

Just as he deviated from his path, a sponge fell at his feet.—A workman was walking by the side of my colleague.

Another at work on the top of the ladder called out to his *confrère* to throw him the sponge.

The workman below picked up the sponge, and then looked up at his comrade, who was perched on the top of the high ladder, and busied with the shutters of the second storey...

Perceiving which, he who had picked up the sponge—WETTED IT IN THE GUTTER,

—and *thus rendered it sufficiently heavy* to be thrown to its destination...

II.

What can we add more eloquent than this act of the brave workman, who thus does more, and proves that he therefore knows more, than all the Academies together?

Is it necessary to call to mind that of three balls of equal diameter, thrown with equal force against the slightest current:

—that in lead reaches farther than your eyes can perceive it;

—that in cork stops within a few feet of you;

—and that in pith returns to you.

And if the bottom of a shuttlecock were not heavy, the shuttlecock would not cut through the air:—it would remain attached to the battledore.*

---

* J. A. Sanson, "Du Point d'Appui."

Not only do we believe that it is rigorously necessary to be specifically *heavier*, more *dense*, than the air to master the air; but I am personally convinced that the *heavier* we are the *better* we shall travel in the air.

And I already know that I am no longer the only one of this opinion.

Where are your proofs? it may be asked.

Take a sparrow or a pigeon with you into the car of a balloon, as I have done many times, and when you have ascended some hundreds of yards, open the cage:

Elsewhere so timid of man and so prompt in flying from his presence, the bird will now remain motionless at the opened door;

Because it feels perfectly well that at this height the air is not sufficiently dense to support its flight, and that the assistance of its weak wings could not protect it against the law of its weight.

Place it upon the edge of the car:—Impressed with a sensation of fear on finding itself above the space which is no longer its own, it will be seized with giddiness—the *vertigo of birds**—and whirl itself awkwardly towards the centre of the car, under your very feet.

Throw the bird out:——You will see it either fall like lead, or turn over and over till in its descent it reaches the denser portion of the atmosphere, where its small size permits it to sustain itself and to re-employ its ordinary movement.

However, proud and alone, the eagle soars easily to the elevated regions it frequents with one effort, by reason of the great width of its wings, which are in proportion to the weight of its body. And the South American condor sails through the air at an immense height before reaching the summits of the Cordilleras.[†]

And why?

---

* "Mémoires du Géant," 2d. ed.: E. Dentu, Publisher.

† "Humboldt has seen the condor rise to a height of eight thousand yards. There is nothing to prove that it cannot rise to a greater height."—D'Esterno, "Du Vol des Oiseaux" Paris, 1864.

——Because of all the flyers properly so called, they are the largest, the thickest, and consequently the *heaviest*.

III.

So admirable is the order of Nature, that all things hold together by imprescriptible links. Where the most important missions are confided to servants apparently the least qualified, ——where the labour of the most miserable insect is as serious as the force which governs worlds, ——nothing is indifferent. There is no detail so imperceptible that it does not cry out to the attentive ear to claim its existence indispensable to the immeasurable *ensemble*;—— there is no sigh so humble or so furtive, the absence of which would not create discord in the wonderful symphony of the Universe.

Thus it is that, when you have fallen upon a truth, everything announces, everything proclaims, everything imposes this truth upon you.

You can no longer open your eyes, or take a step, without the truth starting up everywhere, and without feeling astonished and gratefully delighted in the presence of this incessant and infinite concordance of the proofs of the Eternal Logic.

These proofs follow you, attach themselves to you, and if your idle or distracted attention overlooks them, or does not take them into consideration, they return again and again to the attack, obstinate to the rescue. They wait for you, they watch for you, hidden behind this corner of a wall, or that turn of a path,—even in your dreams they will pursue you...

A few months ago, I was again crossing the plains of Holland, lulled in the somnolent torpor of travel, and vaguely contemplating the interminable green ribbon that unrolled itself beneath my gaze through the window of the carriage, —a ribbon that appeared to be striped with silver by innumerable streamlets, and diapered by thousands of those fine and peaceful cattle that feed for ever on the inexhaustible grass...

Here and there, white and black, like the Prussian escutcheon, some storks, in quest of frogs and salamanders, seemed to be gravely measuring the pasturage with their red compasses...

I started, awakened as it were by a sudden reflection.

—These same storks, that I had so often left behind me without deigning to listen to them; these storks, whose indulgence had not despaired of me, who were waiting for me this time as they had waited before and as they would wait again;—these honest and wise storks also threw to me, on my passage, their certificate in favour of the principle *Heavier than the air*...

Do we not there again encounter the law of truth? Do not the migratory birds, in fact, make up for the insufficiency of each taken isolatedly by the scientific disposition of their masses in the horizontal and vertical arrangements of their well-organised flights?

Do not we Parisians witness every autumn the long and noisy convocation, the *mise en train*, and the final departure of the *ensemble* of our swallows, who have ironically chosen the Dome of the Institute for their starting-point: —beautiful sailers, nevertheless, the brave little creatures! but who understand the advantages of association quite differently and much better than our poor human intelligence, perverted by a deplorable and eternal spirit of antagonism.

And again, how would the heavy quails, with their short wings, succeed in crossing the Mediterranean, if they did not oppose to the stormy winds the resistance offered by the mass of their migratory multitudes?

Have we not here the secret of those few despairing creatures, the isolated laggers-behind, sometimes met with by the astonished sportsman when they have imprudently failed to respond to the signal of the general departure.

And who have incurred and undergo the punishment—a sad punishment indeed!—for having neglected to join in the coalition!

I V.

I have said that the truth pursues us.

At the time I was engaged in writing this page, one of our most eminent colleagues, the most patient and the most conscientious of observers, M. L. de Lucy, called to make me acquainted with some tables drawn up by him a few months since, for the purpose of determining the proportional relations between the weight of all the flying creatures whatever, and the diameter of their extension.

From the largest birds down to the most weakly coleoptera, from the vulture down to the woodpecker, M. de Lucy had measured and weighed them all with that minute attention and scrupulous accuracy that characterise him. The result of these slow and patient investigations is entirely in favour of the law of *Heavier, much heavier than the air,*—a law unknown yesterday, dogmatical to-day, and which upsets most singularly all the suppositions hitherto accepted, for it compels us to admit that:

—*The less weighs* the animal destined to fly, *the wider relatively is the surface of its wings*, or of the plan which it opposes to the air;

And reciprocally,

The greater the weight of the flying animal, *the less considerable* is its relative extension. And however paradoxical this law may appear at first sight, I may add that, of the immense quantity of observations which establish it, not one has been found to disagree with the inflexible rule of these invariable proportions.

For the complete elucidation of the matter, it would be necessary to ascertain the muscular power employed by each of these species in its flight,—an appreciation that it is almost impossible to obtain.

But, whatever we may have still to learn in this respect, how far are we already from those ultra-Academical calculations, which, proceeding by the A + B of the goose and of the pigeon, did not permit of the possibility of a man sustaining himself in the air unless provided with wings having at least an extended surface of *twelve thousand feet*!

V.

However, I still hear some persons persist in accusing us of the audacious, the mad pretension of upsetting all the laws of Nature,— I mean those verified and stamped by the Academicians.

To set aside at one sweep Montgolfier's magnificent discovery instead of applying ourselves to its improvement, and to attempt at one bound the most unreasonable of struggles against the first and the inexorable law of the weights of bodies,—what impertinence!

The problem of Aerial Navigation is in fact double; and if our theory of *Heavier than the air* be hypothetically and for an instant admitted, so far as it concerns the *direction* properly speaking, it is certain that before the *direction* we ought to commence by the ablation or ascension. If we require to be *heavy* to direct our movement, it seems indispensable at the outset that we should be *light* in order to rise.

Now certain *savants* well up in their tables of logarithms object: "It is acknowledged on the one hand, that a body falls sixteen feet in the first second.

"On the other hand, it is proved that the maximum force of man could not enable him to rise more than three feet in a second.

"Therefore, subtracting 3 for the rise from 16 for the fall, there will remain 13; and "consequently the prohibition for man to fly!"

"As regards locomotion in the midst of the waters," add others of the same learned body, "Creation has arrived at *sufficiently* gigantic proportions in *giving* us the whale; but in the matter of aerial locomotion, Providence *stopped*, and *forcibly*, at the eagle and the condor.——In the same manner, by afflicting him with his weight, the same Providence notified as an ultimatum to man that the ground was to be for ever his veritable and invariable place."

According to these reasoners, who hold to their creed with remarkable coolness, the ostrich must have been forbidden to fly;——the pterodactyls and the epiornis, long since dead and buried, exist no more to reply to them.

——And, in order still further to prohibit man from exceeding certain proportions of Nature, let us piously affirm that the *Great Eastern* is less voluminous than the whale;—— let us ordain that the horse exceeds in speed and size the locomotive and its train of carriages;—let us decree that the electric telegraph reaches less far than the human voice, and that the eye of the fantastic lynx is optically stronger than the telescope (engineers! hats off before the aureola of the castor!);—and let us at once stop up the tunnel through Mount Cenis out of deference for the rabbit-hole.

For the requirements of the cause they advocate, they have forgotten their usual theme,—Universal Order created entirely for the wants and for the satisfaction of mankind. Impious blasphemers of the Deity, whom they measure by their own rod, insolent towards the Creator and the created, they would at present deny the miraculous intelligence given to God's image, by the aid of which Man has triumphed in every order of the faculties possessed by animals, in proportion as he has *willed* and *merited* to triumph.

What, then, Man — speedier than the stag, prompter than sound, equally at home in the domain of the fish and of the mole, and who has acquired dominion, even to slavery, over the impalpability called light,—Man, this favourite of Providence, this image of the Divinity, should not rise in the air, like the miserable caterpillar of yesterday, or like the filthy fly born of corruption!*

VI.

The Law of Weight!

But, O Academicians! you who have determined, regulated, fixed, and decreed the Law of the Weight of Bodies, now that your very learned work is finished, will you have the goodness to tell us *what Weight is*?

—How much does the boy's kite in the clouds weigh?

—How much does the soaring bird weigh?

—How much does the flying arrow weigh?

—A skater of ordinary weight, say ten stone, is supported by ice of from two to four inches'

---

* "Mémoires de Géant."

thickness. Gliding along with all the celerity due to the force of impulsion, he arrives at a crevice where the ice is not more than half an inch thick, and passes over it again and again without having cracked it.——How much do the ten stone of our skater weigh upon the crevice?

——How often, in America especially, have suspension bridges been seen to give way through a train stationing upon them,——bridges which had till then supported without budging the passage of thousands of fast trains heavily laden?

——Four men are engaged in letting out slowly a bent cable, to the end of which is attached an enormous mass of stone; the said stone, lowered gradually from the roof, arrives at last, by almost imperceptible movements, upon the ground.

But let one of the four men neglect for a second the regularity of his movements, and let go too quickly an inch of his share of the cordage: with the promptitude of lightning the mass, perfectly balanced just before, will be suddenly precipitated, and will draw down with it the four men, their scaffolding, and even the wall which was the support of the entire operation.

—How much did the stone weigh in the former case? How much did it weigh in the latter? And again, in the interval between the one and the other case, how much did it weigh?

—Upon what basis does your weight agree with the science of Ballistics,—and at what point in its traject does a cannon-ball commence to become fatigued ?—At what other instant will it no longer be able to resist the want of repose?

—Why do not the stars fall?—And by what right does a cloud retain till a given moment its charge of snow, or rain, or hail?

—Why?

—Because, like all other physical laws, the law of weight is not an absolute but a relative law.

Because, according to the different circumstances of its constitution, its action, and its medium, the same body that weighs at one moment does not weigh at another moment.

Because, in combining,* two different forces — weight and projection, for instance—produce compound effects, effects so much the more variable *ad infinitum*, that the forces in question are called upon to act in a medium capricious itself in its degrees of resistance or assistance;—and therein only do we think it probable to solve the mysteries of the immense problem of Aerial Navigation.

—*"The weight of man is the eternal obstacle to the flight of man,"* says the Academy; *"and against this weight the force of man must for ever remain powerless!"*

But the Academy ought to tell us first of all what is the maximum it has decreed for human force.

In our opinion, human force is human intelligence, and that force cannot certainly be estimated by avoirdupois or any other weight.

---

* "Whatever may be the intensity of any absolute force, the slightest other force applied to it angularly, immediately gives birth to a new force, foreign to the first two as to direction, and often also as to intensity… In aerial locomotion, there are the *primary*, the *composing*, and the *resultant* forces."—(Michel Loup, "Solution du Problème de la Locomotion Aérienne" Paris and Lyons, 1853.)

It is this true force that enables me to turn and re-turn a cubic block of granite, weighing a couple of tons, on all its sides,——a block which, without this force, and limited to the simple employment of my muscles, I should not be able even to shake.

This same force enables the servant, who could scarcely carry two pailfuls of water for several minutes, to raise in the body of a pump, per second, and during one hour, a column of water measuring more than thirty-two feet.

Fix limits to, and place obstacles in the way of human muscular force: human intelligence upsets your obstacles and sets aside your limits almost without an effort.

The Ancient Hercules was a man in all the vigour of mature age, with salient and elastic muscles.

The Modern Hercules is a child leaning upon a lever.*

---

* Louis de Lucy, "Le Problème de l'Aéromotion" (L'Aéronaute, No. 4).

VII.

Who would be bold enough to swear, in this age of scientific miracles, that man's weight ("the eternal obstacle to the flight of man," by order of the Academy), will not be the very weight turned to account by man himself as his primary motive power for raising himself in the air?*

—"Nothing is easier than that which was accomplished yesterday," said Biot; "and nothing is more difficult than that which will be accomplished to-morrow."

VIII.

But, if we wish to arrive at any result whatever, let us be upon our guard against acquired formulae; let

---

* Just as I am writing these lines, an Italian gentleman requests the Institute to answer several Papers he had addressed to that learned body more than twenty years ago! —In one of those Papers the author proposed and developed the *means for employing the weight of the human body as a motive agent*.

us closely examine all recognised traditions; and especially, men of good will, my friends! let us stop up our ears as we pass the Institute!

It might too readily be suggested to me, that I have my reasons for avoiding the *savants*.— Agreed.

But all the respect which I owe to Official Science will not prevent me from examining the archives of its proceedings, the endless Martyrology of Truth and Common Sense.

There is no necessity for us to go back so far as Galileo, and we will also leave Salomon de Caus out of the question for the present.

But did not Official Science decree that the attempt to cross the Atlantic with steam was equivalent to the idea of travelling to the moon?

—that the wheels of locomotives would always slide along the rails without advancing ?

—that, moreover, the speed of traction would be sure to stifle the travellers?

—that lighting by gas was an absurdity and an impossibility, &c.?

We might almost affirm that, in the long list of discoveries due to the exercise of human intelligence, there is not one but what was at first denied and opposed by the Scribes and Pharisees of Science.

Is it necessary to call to mind such discoveries as vitrification, the loadstone, phosphorus, gunpowder, spectacles, the microscope, &c.,—all due, as well as a multitude of others, to ignorant hazard alone,—to prove that if Official Science is opposed to those who wish to run, it is because Official Science itself is only capable of walking.

In truth, when, with the History of the Past before me, I contemplate the great scientific game played by humanity ever since the world has been a world, I am obliged to recognise this fact:

—That as the game has advanced, the *savants* are really unrivalled in the art of making the points:

—But it was not they who played the game!...

I firmly believe that Official Science never heaped so many heresies and absurdities as those we are about to meet with against the possibility of Rational Aero-locomotion.

We may state first of all, that at the commencement of the Montgolfiers the Academy of Paris did not hesitate in encouraging the researches made with the view to the *direction of balloons* properly speaking, which eighty years of vain attempts have now proved to be perfectly chimerical. The Academies of Lyons and of Dijon (I do not know what was said by the others) also joined their deceptive encouragements to those of the Academy of Paris.

By a perfectly logical contradiction, the Academy, to keep up its natural consequence in the absurd, ought necessarily to deny the possibility of the solution, immediately the problem of Aviation was placed upon its veritable terms in the name of *Heavier than the air*. This is the course it has at present taken.

Let us therefore examine the arguments opposed by Official Science to the theory of Mechanical Flight.

IX.

Let the Academicians, let Bacon, and Leibnitz, and Buffon, and Cuvier, and even the Encyclopaedia, persist in maintaining that the bird (and consequently also the butterfly) rises in the air, like the Montgolfier, by virtue of the hot air it contains.

(By the way, Buffon, the great diviner, penned the phrase: *To sustain itself in the air, the bird has only to flutter its wings feebly.*)[*]

First of all, Borelli[†] agrees with the Dutchman Nieuwentyt[‡] in calculating that the effort spent by the bird in its flight exceeds *ten thousand* times the weight of its body;—in other terms, he attributes to the muscles of the wings of a goose weighing six pounds an effort of sixty thousand pounds, that is to say, the force of *five thousand men*, or of *five hundred horses*—whichever you like!

---

[*] "Discours sur la Nature des Oiseaux."

[†] "De Motu Animalium."

[‡] "Merveilles de la Nature," 1715.

Then we have the Academician Coulomb (1780), who, for sustaining a man in the air, demands wings of *twelve thousand feet* surface, one hundred and eighty feet long.

—Although a goose weighs more pounds than it has square feet of wing surface; and although each wing of a windmill of two hundred square feet surface produces an effort far superior to the weight of a man.[*]

Then comes the ardent Lalande, who, not satisfied with endorsing the twelve thousand feet of wing demanded by Coulomb, estimates that, for the requirements of certain movements, it would be necessary to double, if not to triple that unlikely surface.

"Therefore," concludes Lalande, "the impossibility of keeping up by striking the air is as certain as the impossibility of bodies void of air rising by their specific weight."

---

[*] Dubochet.

And, exactly one year afterwards, month for month, the ascension of the first Montgolfier gave to these cool assertions a first and a brilliant contradiction,—while waiting for the second.

And I find this other phrase in a letter (dated 28 Messidor, an I.) addressed by the same Lalande to Garnerin concerning an ascension made by the said Garnerin:—

—" ...Why was I not of the party?"...

Always the same story: "*I am a bird*, look at my wings!"*

Then again we have a Report presented to the Academy, and signed by such names as those of Condorcet, Monge, and Bossut, if you please! —which Report re-affirms the demand for *twelve thousand feet* of wings.

And the Academy unanimously approves this Report, "*intending to prevent the undertaking of foolish and dangerous attempts.*"

---

* "Je suis oiseau, voyez mes ailes!"—*Fables de Lafontaine.*

But all these assertions of the *savants* were in vain: the ignorant truthseekers, unwilling to abandon the flag of Mechanical Aerial Navigation, persevered in returning to the charge.

It was then that the Academy, in order to finish with the question once for all, hurled at the obstinate theorists one of its best calculators, Navier, who drew up a new Report (dated September 6, 1830), no less affirmative, no less prohibitory, no less approved of by the unanimous Academy;— and it is with this Report, this Act of Faith, this Academical Credo, that our opponents endeavour to crush us at present.

—*Magister dixit.*

Let us give a specimen of the Absurd carried to the Fantastic. Listen to the Report:

"Man," affirms doctorally the learned Navier, "Man does not possess, all proportions taken into consideration, the NINETY-SECOND part of the strength spent by a bird to sustain itself in the air (!)

"When a bird hovers, it beats its wings *about* TWENTY-THREE times in a second (!!)

"When it hovers, the quantity of action spent by the bird in a second is equal to that which, it would require to raise its own weight twenty-four feet (!!!)

"When it moves horizontally, the quantity is equal to that which it would require to raise its own weight *about* FOUR HUNDRED YARDS (!!!!)

"The quantity of action is *so much the less*, for the rapid flight, that the density of the air is not so great." (!!!!!)

And, I repeat, the Academy unanimously approves and countersigns such extravagant assertions!

X.

The Folly of Figures!

Apply these terrible calculations of Navier to the salmon for example, that the thinnest line stops, and behold the salmon called upon to employ for its ordinary circulation a force equal to that of a steam-engine of *fifty-horse power!*

But, it may be observed, the medium is no longer the same.

That may be. So let us remain with the honourable M. Navier; and thereanent he has forgotten one point, which is, that the learned calculations heaped up by him to deprive man of his right to fly should also prevent birds from flying, since they require from a goose *the force of four men* for its slowest flight.

Touching advantage of our sweet ignorance! We simpletons, who do not seek the miraculous, have only to raise our eyes to perceive, without the aid of any algebraical calculations whatever, that—as our children say in playing—

OUR BIRD FLIES!!!

Some birds fly a whole day. If, by Academical order, we were to admit the figures of the mathematician Navier, what an incalculable outlay of force must have been incurred by such birds when evening arrives; and how is it possible for them to have supported such fatigue?

The bird *Frigate*, which is not a web-footed bird, is encountered at a distance of four hundred leagues from land.

Four hundred leagues to come and four hundred leagues to return, make eight hundred leagues, without reckoning the detours and the deviations.

How many millions of pounds has this Frigate therefore raised since the morning, according to your figures, O good M. Navier?

Other birds, provided only with the least perfect appliances, — the quail, for instance, — travel several hundred leagues in a single flight; and this incommensurable effort is accomplished without even having experienced the want of repairing their strength, without repose, without eating, and without drinking!...

How then has man been so blind or so stupid hitherto as not to have used for industrial purposes such reservoirs of inexhaustible force?

But let us pass on to the consideration of serious questions.

XI.

Of all the Physical Laws, there is not one which has been so little and so imperfectly studied down to the present time as that of the flight of birds and insects.

—"We speak of the muscular power of birds," wrote Captain A. Girard, in 1858: "the cause which enables them to sustain themselves in the air is as completely ignored as was, before Pascal, that of the rising of water in the body of a pump, which used to be explained by Nature's horror of a vacuum."

Quite contrarily to the supposed and formidable calculations we have just cited, we consider that, of all the means of locomotion, flight is perhaps that which requires the smallest outlay of force.

When we examine the relative weakness of sea-birds, which, better than all others, soar with perfect ease in the midst of the most violent tempests, it appears to us that the strength of the pectoral muscles has been over-exaggerated by certain pre-oc-cupations which prevented the observation from being accurate.

This pretended exorbitance of strength is as fanciful as the idea of ascension by means of specific lightness, which was so long and so obstinately maintained.

And if Nature has endowed certain birds with considerable pectoral development not met with in other flying species, it was not for the requirements of flight properly speaking, which is not at all pro-duced *solely* by a dynamical cause; the pectoral development in question was given in view of other necessities, such as rising rapidly and vertically in given cases,[*] stopping suddenly the flight, combat-ing and carrying off prey, &c.[†]

If, in fact, the faculty of flying depended also principally upon the strength of those muscles, how should it happen that a grain of lead sent into the extremity of the wing,——a single feather twisted,——nay, still less, and without touching the

---

[*] Michel Loup.
[†] Dubochet.——J. A. Sanson.

feathers, the simple ligature of the feet,— suffices to prevent or to stop the flight of the strongest bird?

There is therefore in the phenomenon of flight something quite different from the simple dynamical question; and that is, a primordial correlation with other elements of interest quite as essential,— statical, mechanical, &c.

All the more or less arbitrary estimates by order of which Official Science has insisted and still insists upon interdicting to man the possibility of imitating the flight of the bird,—all those estimates, we repeat, were they even correct, could not be admitted to have any weight in the cause, because they are isolated. An uncorroborated witness is no witness.[*]

In all justice, that is to say, in all reason, they cannot be accepted without taking into parallel consideration, with an equipollent preoccupation, the other phenomena whose combined action is necessary to flight,—phenomena whose action is at

---

[*] "Testis unus, testis nullus."

least equal, as we have said, perhaps superior, to the dynamical effect which Official Science has hitherto obstinately persisted in alone considering.

Now, in the phenomenon of flight, we have already acquired a primary truth of the highest import.

It is beyond doubt, in fact, that the weight of the flyer is regulated by each movement of the wings, and that it is compensated and decomposed in proportion with the force of progression, aided by the different inclinations of the wing-surface.*

Other deductions, not less interesting, may be at once drawn from this primary truth.

——IN THE CONTINUED FLIGHT, THERE IS NO ACCELERATED FALL.

——THE ACTIVE FORCE OF ONE TIME OF MOVEMENT IS PROFITABLE TO THE TIME FOLLOWING.

---

* D'Esterno, "Du Vol des Oiseaux" Paris, 1864. Consult again, and constantly, Michel Loup.

——THE FALL AT THE FIRST MOMENT IS
ALMOST NULL, ESPECIALLY WHEN THE SUS-
PENDED BODY HAS A LARGE SUPERFICIES.

——THE FIRST PRINCIPLE OF CONTINUED
FLIGHT IN THE BIRD IS THAT IT NEVER ALLOWS
ITS FALL TO COMMENCE.

To resume, and consequently:

——All the Academical and ultra-Academical estimates of the force spent by the bird in its flight have
been exaggerated *ad absurdum*.

XII.

From these incontestable data, collected or appre-
ciated long since by such observers as Meerwein,
Huber, Chabrié, J. A. Sanson, senior and junior,
the Swede Tollin, Deniau and Dubochet (of
Nantes), Jourdan, Vignal and Michel Loup (of
Lyons), Cagniard de Latour, Franchot, Babinet,
J. A. Barral, Beleguic, Emm. Liais, de Semallé,
d'Esterno, Pline, de la Landelle, Taillepied de

la Garenne, Vaussin-Chardanne, L. de Lucy, Arwed Salives, Arth. Gandillot, Jules Verne, Bourcart, F. du Temple, Henry Bright, Alph. Moreau, Preslier, Laubereau, Jullienne, Julien (of Villejuif), Lord Carlingford, Parisel, Briant du Lescoët, L. Grandeau, A. Sanson, Th. Maurand, Serres, Piallat, Richard, Camille Vert, de Ponton d'Amécourt, Telescheff, George Cayley, Panafieu, Landur, Tigé, Steiner, Diesbach, Garapon, Engel, de Louvrié, Mareschal, Danduran, Doctor Juge, Doctor Wolf, Doctor Tavernier, Vandal, Brochon, Besnard, O'Frion, Perrée, Guyot, Davelouis, Lacan, Brisson, Mure, de Montdesir, Van Monckooven——(I ask pardon of those whose names I forget),——it would seem to result that there is some chance of the possibility of man rising, sustaining himself, and directing his movements in the air in imitation of the flying animals.

Or, at any rate, that the study of this grand question cannot be considered as an illusory folly.

As to ourselves, personally, we consider the question of human flight as solved from the mere fact of its having been posed.

For whenever man, for the satisfaction of his wants, has sought to imitate Nature, he has equalled, and often surpassed, his model.

He did not possess the four swift legs of the horse, the stag, or the greyhound: — yet with the locomotive he has completely distanced the greyhound, the stag, and the horse.

He did not possess a natatory apparatus like the fish:—yet he goes not only upon the water, but, like the fish, under the water. — He was weak, but he armed himself, and the most formidable animals fly before him.—He has subdued the very flames themselves, and commanded them to carry him.

The empire of the world was really given to man. He conquered all, as soon as he determined upon conquering. When it pleased him to realise the prodigy of being more rapid than sound, electricity bore his words from one pole to the other with the speed of lightning.

When he shall have determined upon doing so, man will fly like the bird, better than the bird; for, without entering here into abstract details, it is certain to us that man will be obliged to fly better than the bird, in order to fly merely as well.

The means for accomplishing this triumph will not be wanting; for if what has been said be true, that a question well posed is on that account already half-solved, the hour has come for the realisation of the grandest of all human conquests.

The continued observation of natural phenomena indicates henceforth to man the rational and certain course he has to pursue.

The Titanic strength of birds, which used to intimidate our *savants*, is centupled by the intelligent hand of man, more powerful than any other force.

Mechanical art, which already in the course of the last century produced Speaking Heads, and Ducks* that imitated even the most mysterious of Nature's acts and movements,— that same art now places at his disposal more means than are required for passing from the theory to the practice of Aerial Navigation.

In 1784 Launoy and Bienvenu exposed to public curiosity a little mechanical helicoptera that rose automatically in the air with its motive power, a simple clockspring with a trigger, wherein the

---

* Those of Vaucanson.

force was, it must be acknowledged, previously laid in store. This was, we believe, the first practical demonstration of the future rational aero-navigation; but the interesting experiment was no sooner exhibited than forgotten.

In 1849 the Lyonnese Vignal re-invented and exhibited a similar model with as little success.

A few years later, in 1853, MM. de la Landelle and Ponton d'Amécourt—and, almost at the same time, Mr. Bright, a very distinguished English engineer—sent up and developed in the air a like description of apparatus.

On witnessing these mere toys rise automatically in the air, the illustrious *savant* M. Babinet had the courage (Academician against the Academy!) to place himself under our flag, and to exclaim:

—"THE DISCUSSION IS CLOSED! IT IS NOW ONLY AN AFFAIR OF TECHNOLOGY AND OF MONEY!"

And, entering at once upon the disputed question of the first possibility of the ascension of heavy bodies, he supported and confirmed, with all the weight of his great authority, this theory, still so paradoxical in appearance, but which we have never hesitated in advancing; namely, that *the more powerful the apparatus employed for mechanical flight, the lighter relatively will it become.*

Beyond the fact that a large machine is always more efficacious than a small one, it is evident that one force of two-horse power weighs less than two forces of one-horse power each; and the higher we raise our proportions, the more we recognise that our weight diminishes relatively as our force increases. The progression of our lightness will therefore augment proportionately with our increase of power, and, far from opposing an obstacle, our amplification will, on the contrary, be the very means of guaranteeing our success.

——"SINCE YOU HAVE SUCCEEDED IN RAISING A MOUSE," writes our learned master Babinet, "YOU WILL HAVE MUCH LESS DIFFICULTY IN RAISING AN ELEPHANT."

The means of *Ablation* or ascension once acquired, the question of *Direction* is merely one of secondary consideration.

Without speaking of mechanical means, the simple theory of the parachute—of the parachute really directed by the aeronaut—suffices to prove that a body falls on the side to which it inclines; so that, the elevation once obtained, the aerial navigator has *effectively* placed at his command a capital of force of which he can afterwards dispose as he may think proper.

XIII.

Let us remark, moreover, and from another point of view, that Nature has only given to Man the aspirations which are in accordance with his aptitudes.

"The attractions," said Fournier, "are proportioned to the destinies."

Now, from Icarus, and the Scythian Abaris (mentioned by Diodorus)—for the principle of *Heavier than the air* has had its heroic period—man has never ceased to raise his eyes towards the dominion of the air, which he feels to belong to him. Everywhere, and always, man has sought the means for travelling in the air: — but never was there a human brain sufficiently "ardent" to occupy itself with a project for living in the fire.

There is no one who has not thought of passing freely through space, both before and since Lana, who, preceding Montgolfier by the invention of his globes void of air, called this aeromania *the Reasonable Folly* (*sapientem stultitiam*). There is not one of us whose aspirations have not indulged in the ecstasy of the immensities without horizon, and who has not dreamt, with our great poet Victor Hugo, of

"La liberté dans la lumière!"

There is no city without its more or less vague tradition of the flying man, from Simon the magician, who rose in the air at Rome in the presence of Nero, and the Saracen who accomplished the same feat before the Emperor Comnenus at Constantinople, — from Dante of Perouse and the flying monk of Lisbon, down to the Viennese Deghen and the Parisian Marquis de Bacqueville.

Most of these attempts failed, others partly succeeded, and a few met with complete success and were even happily repeated.

All these attempts were witnessed by dense crowds of people, predisposed to jeer, because man's vanity is always ticklish on the score of being exposed to deception; notwithstanding, they felt themselves irresistibly attracted by the profound and instinctive sentiment of the Possible, as well as by the secret and insuppressible feeling of Hope.

How is it then that, in spite of this universal pre-occupation at every period of the world's progress, in spite of this unwearied expectation, in spite of these oft-repeated and sometimes successful attempts, and in the presence of a possibility so rationally established,—how is it that the problem of Aerial Navigation has not long since been solved?

—Because nothing comes before its appointed hour, and each hatching process has its duration;

—Because it has been written, that everything shall be reckoned unto man with nicety; because it is ordained that everything he is to enjoy must be first purchased and paid for according to the inexorable tariff; because *If thou desirest anything, thou must first of all merit it*;

—Because every human conquest is paid for with sweat, with tears, and with blood; and that the greater the conquest, the more costly and the more difficult is the payment;

—Because he who precedes others on the way gets his feet torn by the stones and briars of the new road,—and stops short;

—Because the eternal mistrust which man has of himself causes him to repel with hatred and contempt, instead of welcoming, the labourer in Nature's domain who suggests a new or a useful Word.

—Because—but let us stop here, although many other reasons might be adduced.

XIV.

To confine ourselves to the special question under consideration, was there ever a more multiple or a more complex one?

This great Unknown Quantity, which, once set at liberty, will bring about the most gigantic, the most incalculable revolutions in all human relations,—this great Unknown bears at one and *the same time* upon all the physical sciences, for it proceeds from them all.

And this primary multiplicity of the sciences essential to or participating in its germination, is also one of the causes which have so fatally retarded the fatidic hour.—Where, then, is the single brain in the world that will offer us the synthetical universality of the knowledge and aptitudes thus called for?

Scarcely are we agreed upon the principle, upon the point of departure, and behold, before taking the first step, we are astonished and stupefied by the embarrassing and infinite diversity of the means of flight with which Nature has endowed insects merely.—Which of these systems ought we to study the first as the most probable, the most assimilable to our relative faculties?

To this first objective embarrassment must be added the inherent deficiency of the human mind, which, in every invention, never fails to proceed from the compound to arrive at the simple.—What researches, what combinations, what hypotheses, were made, attempted, and considered before Christopher Columbus' egg was broken!

We have everything to do, everything to learn. Nothing remains to us even of those legends of the flying men, who, moreover, only offered us the embryo of that Aerial Navigation which we desire to realise in the proportions or the positive *practical* utility of which it is capable.

Where are we to begin? Who is to guide or to advise us? Where, in this case, is the Professor? For our purpose, all acquired and consecrated Science offers us *nothing!*—And all that the Academicians have taught us is reduced to *zero*!

For the sake of giving an example, we will take the screw-propeller, which, according to all probability, is destined to play a certain part in this matter. "The screw-propeller," says M. Maurice Saint-Aguet, "had been discovered eight or ten times since Dallery's patent in 1793 (and we may add that he was not its original inventor), before it was employed, more than thirty years afterwards by two engineers, one an Englishman and the other a Frenchman, who still continue to dispute very seriously the priority of its application."

We will leave them to their dispute; but let us hasten to admit, as a beginning, that the screw-propeller—which has now been so long in use, which has been studied, tried, regulated, and verified in all its applications as well as in all its modifications—is perfectly known to us; we hold it firmly in our possession; it is our own.—We are quite sure of that, is it not?

Well, a few months since, in a comparative trial before Cherbourg, the *Solferino* was started at full speed, and at such speed that, of the four invariable, indispensable regulation paddles, two were broken.

What was the result of the unforeseen suppression of those two paddles?

——THE INCREASE BY NEARLY A THIRD IN THE SPEED!

Some time afterwards, a second and a similar trial completely confirmed the result of the first.

How then, without making an inventory, can we accept any of the assertions of your *savants* as orthodox?

Therefore, and to resume, if we wish to realise the grand and beautiful dream of the human flight, we must first of all try to discover a mind of the highest order, and universal, having begun by unlearning all that it had learnt from others, and relying wholly upon its own intelligence and experience.

To this complete and perfect *savant* of *ours*, we must afterwards give the Genius which alone makes effects spring out of causes.

Let us then inspire him with restless curiosity, panting research, and the noble and insatiable thirst for great things.

To which let us add, resolute obstinacy of will, absolute self-denial, and unreserved devotion, not to speak, of course, as necessary adjuncts, of the various physical aptitudes, of fundamental courage and of indispensable good health.

And finally, as inventions are ruinous, and as money is dear, and the wood of the bed of Bernard de Palissy is constantly burning, let us slip a few millions, less or more, into the pocket of this *rara avis*,

——and then Aerial Locomotion will become an accomplished fact!

XV.

In the meantime, hours disappear, days fly past, and ages accumulate, while indolent Humanity, with eyes upraised towards the regions it contents itself with dreaming about, appears to be waiting for the solution of this great Problem,——the Divine solution which will suppress frontiers, render war impossible, and tear up even to the last leaf of the different Codes of our barbarous periods, to dictate another and a last one, the Law of Liberty and of Love.

From time to time, it is true, from one point or another, an isolated aspiration is breathed out, a light breaks forth, an effort is attempted, only to be little heeded, less regarded, and soon entirely forgotten!

With this sad conviction, and finding that the predestined being we have been waiting for so long to open to us the free career in the immensities of space did not make his appearance, I thought it was time to provide a centre for all those isolations, and to create a temple for the reception of the communicants.

And, in the absence of some one more worthy, I resolved upon establishing a point of concentration, of comparative examination, and of cohesion for all the efforts, isolated hitherto, and consequently lost.

Appealing to all the earnest investigators in quest of the grand discovery, this free Association, disinterested in everything except the good of the

Cause, would receive the contributions of each in order to make them, little by little, the treasure of all.——Let there be a sanctum where the inventor, the reviled of all times and of all people, would be sure to meet with a hospitable threshold and willing listeners !

So I had the honour of founding our *International Society for the Encouragement of Aerial Locomotion by means of Apparatus heavier than the air.*

But the simple foundation of such a Society did not suffice.

Our Society of Encouragement was, as may well be thought, neither commercial nor civil; and the modest minimum of six francs a year paid by its members, did not constitute a very exorbitant resource.

It was necessary to animate it, to give it the means of growing, of propagating itself, of *serving*. It was necessary, beyond this primary action of concentration, that our Society of Encouragement should complete itself by *efficacious encouragement* offered to inventor ;——it was necessary, after having conceived, that the Society should produce.

But where was the money requisite for the creation of the first attempts to come from?

If I had thought proper, as I was ardently advised, to open a subscription by way of asking the public for forty or fifty thousand pounds, in order to enable me to *try* to make a machine that might *perhaps* ENDEAVOUR to fly in the air, the amiable public would, no doubt, have made a very pretty outcry! All those who will not look, all those who cannot see, all those who love their gold beyond everything, would have said—

"This man is decidedly mad! We used to fancy that he was, but now he confesses it. What! he has the impudence to ask us for fifty thousand pounds to— He is an impertinent fellow, who wants to have the laugh of us!"

I should have received insults, which are much less costly than bank-notes; and some persons, especially those who object to putting their hands into their pockets, would at least have rated me as a thief.

What more? After this profound observation by M. de Girardin,—"One of the first expenses with which a government ought to honour itself should be that of a few millions of francs destined to encourage the researches bearing upon Aerial Navigation,"——could I address myself to the government?

First of all, I do not know how money is asked for from a government; and, moreover, I should not know how to thank it. I think I may confess this defect without inconvenience, for I have been told that it is not contagious.

XVI.

I therefore decided upon earning the money necessary for the constitution of our first trial capital myself. It appeared to me that, as soon as the debt of a few tens of thousands of francs which I had taken upon myself was paid, I might be permitted to make an appeal for assistance to all in the grand interest of all.

I well know the eternal, the insatiable eagerness with which the public flocks to any aerostatic spectacle; and I said to myself that, in order to realise the conquest of the air with apparatus *heavier than the air*, and to kill the balloons which have caused us to waste upon a false scent the last eighty years (the richest in respect to acquired science), I would make a balloon—*the last balloon!*—of such extraordinary proportions, that it would realise what has never been but a dream of the American newspapers; a balloon more than half as high as St. Paul's Cathedral, — a balloon which, in its wicker house of two storeys, would carry thirty-five or forty passengers with the ordinary coal gas, and more than a hundred with pure hydrogen gas,— which should be able, owing to the enormous ascensional force represented by the weight of its ballast, to remain two, three, or four days, and as many nights, in the air, and thus accomplish veritable voyages over land and sea.

And, all alone, I rushed headlong into the great undertaking of the GIANT, an undertaking which promised such excellent results, but which has hitherto been only disastrous.

Of a speculative mind, perhaps, but very badly endowed as a speculator, averse to the counsels of prudence, rebellious towards all arithmetical calculations, confiding to a folly, the last three years of my life have been a series of every description of grief, disappointment, and anguish.

I do not speak of slight wounds, of insult, or of sarcasm, although my skin, as a former caricaturist, is naturally more sensitive than another's; and if I confess to take a little pleasure still in ridiculing others, yet I own I do not like to be jeered at myself.

Now, no scientific discovery, no political occurrence, has ever given rise to more quodlibets, both in rhyme and in caricature, than Aerostation. I possess an engraving of the time, representing the ascent of the first Montgolfier balloon. On the top of it may be read:

"Confusion of the sorry jesters!"

—In 1783,——already!

I have sought for the reason of the fecundity of this constant pitiless opposition to Aerostation; and I believe I have discovered it.

Practical aerostation presents in reality no sort of danger; I mean when intelligence presides over it, but not when it is abandoned——as is nearly always the case, alas!——to the ignorance of the mountebanks into whose hands it has fallen since its birth.——"All aerostatic accidents," says J. A. Sanson, "have been owing to ignorance, to impudence, and especially to drunkenness!"

There is, therefore, not the least courage in going up in a balloon, because there can be no courage where there is no danger.

For all that, the practice of aerostatic voyages still continues to be an extraordinary source of fear to many persons,——women excepted, who are always more truly brave than the men.

I imagine, therefore, that those who would not dare to go up in a balloon avenge themselves by an easy system of raillery on a courage that offends and humiliates them.*

If at any meeting one of the speakers shows himself more ironical, more disdainful, more difficult to satisfy than the others with respect to some aerostatic fact, feel his pulse:——assuredly you will find him to be the coward of the assembly.

---

* "Et je tiens pour affront le courage d'autrui."—— VICTOR HUGO.

By touching upon balloons I had let loose upon myself the whole pack of scoffers. I readily got used to their bites, from which blows of a more serious description were soon to distract my attention. Ruined, become the prey of all sorts of parasitism, deceived, betrayed, robbed, and wounded, I passed through a succession of repugnant contacts, bitternesses, anguishes, and the most scandalous iniquities. In a word, I was vanquished.

But I am not ashamed of that, for my past defeat has never for an instant disheartened me, and I have not once doubted of my future victory. I am not displeased even at having been vanquished at the outset: it is good, it is often an honour and a greatness, to be conquered.

I knew what I was about…

Besides, what concerned myself personally mattered little. Had not the great principle of future Aerial Locomotion been established? Had I not created in France and everywhere else, for the advantage of the Cause of Aviation, an Agitation manifested every day by verbal discussion, and in numerous publications,—a salutary and a fruitful

Agitation, that will only be extinguished by the solution of the great problem?

Our *International Society for the Encouragement of Heavier than the Air* is constituted; and, while waiting for the time when the ascensions of the GIANT will provide the primary elements of action, it is investigating, it is studying, it is collecting information and experience.

To our first appeal in the name of *Heavier than the Air*, a few, from different centres and unknown to each other, had promptly responded with timidity and embarrassment. Afterwards, the workmen of the first hour counted their forces; and as their Faith was the same, a spirit of fraternal confidence was at once engendered, and each drew from under his cloak what he had brought for the examination and consideration of his brothers.

A few gentlemen, eminent for their science, their courage, and their devotion, did not hesitate to brave our "French poltroonery," and put themselves at the head of this little group of convinced and resolute minds. The first names on the list of our staff were those of MM. Babinet (of the

Institute), J. A. Barral, Verdonnet, Franchot, and Baron Taylor.

Another man of firm will, and an earnest apostle of the faith he had already preached during many years, undertook the heaviest task, that of practically guiding, day by day and hour by hour, the destinies of our Association; and no one could have been better endowed with the qualities of gravity, energy, and prudence required for the presidence of our nascent Society. Formerly an officer in the navy, and a clever writer, M. G. de la Landelle had abandoned all his avocations in order to devote himself entirely to the Grand Cause; and I regard my meeting with that gentleman as really providential.

Every tree produces but its own fruit. A pear-tree does not bear apples any more than an apple-tree bears pears. To conquer is something, but to keep is all.

I possessed, perhaps, a little of what was required in the way of devotion and active enthusiasm. Neither believing nor loving by halves, I felt at liberty to risk the bread of those belonging to me, and my own life, in my undertaking.

But all that, and more, was wanting to bring this enterprise to a final termination,——a great amount of patience, order, method, and persistence. By the side of the voice that produces the noise, we stood in need of a case to preserve the sound.

Thanks to La Landelle, thanks to the phalanx of *élite* who grouped themselves around him, our Society, so humble at the outset, is now organised, has a standing, and has passed nearly three years in the pursuit of its labours. Engineers, army and navy officers, manufacturers, working men, public functionaries of the highest rank, physicians, chemists, mathematicians, (some Academicians even!), have flocked to us in daily-increasing numbers. Under the constant influence of De la Landelle, all this little world of well-intentioned people has grouped itself into different committees according to the peculiar aptitudes of each. There is the engineers' corner, the arithmeticians' corner, the naturalists' corner; and there is also a goodly muster of men of the world who have been attracted to us by the extreme interest of the researches we have undertaken. Projects are addressed to us from all quarters; for the large class of inventors, generally so distrustful, at once understood that the Insubstantial Being called the Society of Encouragement was a disinterested and a devoted Being.

And thus we pursue our weekly labours, studying and preparing the way till the day arrives when, to facilitate our passage from theory to practice, the GIANT will call upon each of the capitals of the world to pay its share in the ransom of *Heavier than the Air*.

While awaiting this approaching day, and even when it shall have appeared, if, O Reader! you have become convinced, not of our zeal and self-denial, but that we are engaged in the pursuit of an Idea destined to bring about the most generous, the grandest of the revolutions of humanity,—assist our *International Society of Encouragement*, and join your efforts, whatever they may be, to ours, for the propagation of our labours.

If I am dreaming, let me dream on,—but I defy any one to awaken me!—Let me contemplate the air studded with barques travelling with such rapidity as to humiliate the Ocean and all the locomotives of the Earth!

From all the points of the world I see man rising with the promptness of electricity, soaring in the air, and descending like a bird when and where he wishes.

Books relate that people formerly travelled on roads of iron in horrible boxes with intolerable slowness, and exposed to insupportable annoyances... A frightful lacing-motion backwards and forwards shook the traveller from his departure till his arrival; and a dinning chorus of chains, wood, and shivering windows, was the funereal-music accompaniment of those unpleasant trains. During the long journey, the dust entered through the air-holes of those cruel boxes in such quantities as to cover the unfortunate traveller with its stifling winding-sheet.——At that time, a voyage was a fearful trial, not undertaken with anything like cheerfulness.—— Who would believe that man had only to will in order to deserve the aerial routes which now appear to us so charming, and that he preferred suffering for many ages such atrocious torments!...

Those poor people used to think they had made a great progress because they travelled somewhat faster on their roads of iron than in their carriages drawn by horses, which were the beginning of all locomotion. They endeavoured to console themselves with certain statistical returns, which seemed to prove that the number of road accidents was somewhat diminished.——Let it be noted, *en passant*, that they had not even been able to discover the equivalent of our parachutes!

Their statistics were perhaps tolerably correct; but when an accident did happen, what disastrous results!——Hundreds of people crushed, burnt, annihilated, through a mere trifle having been placed across one of their pitiful roads!

How different from our aerial voyages, without shocks, without concussions, and free from noise, dust, fatigue, and danger!

And how is it that the human race has waited during so many ages for deliverance, when, for its redemption, it had only to make an intelligent employment of the first elements of Statics and Mechanics?*

NADAR.
PARIS, March, 1866.

* "Mémoires du Géant."

Biographical Notice of M. Nadar

Félix Nadar, the son of a very worthy Lyonnese bookseller, was born in Paris on the 5th of April, 1820. On leaving school, he was entered as a student at the School of Medicine, and attended the Lectures, sometimes at Paris and sometimes at Lyons, during two years and a half. Being without any patrimony, he was obliged to earn his livelihood at the same time that he pursued his studies, and thus began his career as a newspaper writer at the early age of sixteen.

Young Nadar's taste for letters very soon caused him to abandon the study of medicine; and he became the secretary of one of the members of the Opposition during Louis Philippe's reign. But finding that situation too calm and too quiet for his feverish organisation, he attached himself to the editorial department of *Le Commerce* newspaper, then under the management of M. Charles de Lesseps. He passed five years in this occupation, during which, besides his numerous political contributions to the journal, he found time to publish " La

Mort de Dupuytren," a little chef-d'oeuvre, afterwards translated into every language, and reproduced in so many provincial and foreign papers, that it is said to have had six hundred editions! This success, so encouraging for a young man, was followed by a three-volume novel, entitled "La Robe de Déjanire."

While thus fairly launched on the waves of literature, the Revolution of 1848 broke out, and his lively sympathy with the sufferings of the Poles induced the enthusiastic Nadar to join the Polish Legion: he still continues to regard this action as one of the most reasonable of his life. He and all his companions were, however, arrested and imprisoned, first of all in the casemates of Magdeburg; but they were afterwards removed to another Prussian town, Eisleben, where they were detained several months.

As soon as he was restored to liberty, Nadar returned to France, and threw himself into the political turmoil of journalism, at that time free from its ordinary restrictions. During the agitation which prevailed throughout the Republic on the subject of the nomination of a president, Nadar worked day and night, with his pen and with his pencil, in the *Charivari*, in the *Journal pour Rire*, and in a very celebrated work called *La Revue Comique*, now no longer procurable; and all his talent was employed against the election of Prince Louis Bonaparte.

After the coup-d'état Nadar retired from journalism, the restrictions upon which imposed by the law of 1852 deprived him of the entire liberty which men of his stamp require for wielding that powerful weapon the pen.

At this time he terminated the *Panthéon Nadar*, an immense lithograph (printed upon a single sheet) containing the portraits of all the well-known men in France. From his continual and intimate relations with those who represent what may be called the aristocracy of the mind, no one was more capable of giving the exact "charge" of the two hundred and fifty celebrities, or semi-celebrities, comprised in this artistic production, now become very rare.

At about the same time Nadar founded a noble photographic atélier, which soon became very fashionable, and is still the most important establishment of the kind in Paris. He was the first to apply the electric light to the art of Photography, and published some very curious views of Underground Paris,——sewers, catacombs, &c., taken by this new process.

Nadar had long thought of turning Photography to account for making planispheric or strategical plans, by employing apparatus taken up with him in a balloon. And here let it be mentioned, that Aerostation has always pre-occupied and attracted his attention. The purpose just alluded to presented him with a pretext for ballooning, and with that object in view he made about thirty ascents, the results of which *practically* proved, what he had previously supposed, that the direction of balloons, properly speaking, is a chimerical impossibility. But believing, on the other hand, that man has the power of going everywhere an animal can go, Nadar commenced an extensive agitation in favour of the future *Rational* Aero-navigation by means of apparatus *specifically denser than the air*; gave himself up, heart and soul, to this work of progress; and, indifferent to obloquy and incredulity, boldly attached his name to his project, and founded an International Society for the Encouragement of Studies relating to Aerial Locomotion. As a proof of his zeal and heartiness in the Cause, he set apart a portion of his residence for the meetings of the new Society, of which, moreover, he himself furnished the first pecuniary resources.

Shortly afterwards M. Nadar, setting aside all personal interest, neglecting the management of his photographic establishment, and sacrificing everything to the object he had in view, showed no hesitation (in a country where all initiative is so rare, and where every one is accustomed to rely upon the Government) towards engaging himself, personally and alone, in the outlay of 200,000 francs for the fabrication of the aerostatic monster "THE GIANT." With the receipts he expected to obtain from a certain number of ascents, he hoped to realise an experimental capital for the Society of Encouragement.

The first departure of "THE GIANT" took place from the Champ de Mars, in Paris, on the 4th of August, 1863. The balloon rose with its valve open, in consequence of the oversight of its primary constructors, the Brothers Godard; and it fell, four hours after, at a few leagues' distance from the starting-point.

For "THE GIANT'S" first ascent, Nadar wished to upset a popular superstition attached to the number 13, the number of its passengers, amongst whom was the Princess de la Tour d'Auvergne and Prince Eugène de Sayn-Wittgenstein.

A fortnight after, a second ascent was made from the same place at six o'clock in the evening. During the night, just as the balloon was approaching the German Ocean near Amsterdam, a spirit of revolt was manifested by some of the passengers; and the Brothers Godard insisted upon descending at any cost. Nadar immediately seized the cord commanding the valve with one hand, while in the other, it is said, he held a revolver, and so re-established order "on board;" but the occurrence was of unfavourable omen.

A breeze from the coast then acting upon the balloon, carried it towards Germany, where, after continuing its course wildly for some time, it fell at Nieuburg, in Hanover.

The entire world has been made acquainted with the terrible catastrophe so dramatically related in the "Memoires du Géant." In the space of about fifteen hours, the balloon had travelled over eight hundred miles, the most considerable voyage on record in aerostatic annals. The force of the wind was fifteen leagues an hour; and two of the most important cords soon snapped just like threads. As it was impossible to disinflate "THE GIANT" quickly enough by the orifice of the valve, which had been left too narrow in spite of the reiterated observations made by M. Nadar to the Brothers Godard during the construction of the balloon, it was dragged through a space of seven leagues in thirty minutes, striking against the ground in a series of leaps of from one to thirty yards in height; the car at times scraping over the obstacles it encountered in its flight with the rapidity of a locomotive, while the passengers held on by the cordages during the succession of terrible shocks. Of the nine passengers, one of whom was M. F. de Montgolfier, some fell and others slipped through the openings made by the injuries to the car. The Brothers Godard, forgetting, in the presence of the danger, every feeling of responsibility, were the first to jump out; and the balloon, thus lightened, continued its wild course with fresh violence.

Madame Nadar, yielding to a dark presenti-ment which had taken possession of her mind with regard to this voyage, had accompanied her husband, who, holding on to the cordages and squeezing his wife to his breast, to protect her as much as possible against the shocks, remained in the balloon with his precious burden, till "THE GIANT" at length stopped and fell over them both.

One of the passengers had his arm broken; M. Nadar got his right leg fractured; and Madame Nadar received injuries which confined her for several months to her bed.

A strange circumstance characterised the second ascent of "THE GIANT," at which the Emperor Napoleon, without having announced his intention, unexpectedly assisted. M. Nadar, whose political opinions are very decided, refused for some time to receive his visitor. Some of the English papers gave an account at the time of this singular interview, of which Nadar certainly had not sought the honour,— and of the aeronaut's more than glacial attitude before the present sovereign of the French people, little accustomed to the display of such deliberate coolness in answer to offers of assistance, politely but firmly refused by Nadar. Notwithstanding, it is said that the Emperor takes all the more interest in the trials and courageous attempts of a man whom he cannot do otherwise than respect.

From the interest attached to the adventures of "THE GIANT," the managers of the Crystal Palace were desirous of organising an ascent in London; the wounded condition of its captain rendered the execution of that idea impossible, but as soon as the doctors permitted, M. Nadar accompanied his balloon to Sydenham, where, as will be remembered, it was exhibited during a month.

Since then, three fortunate ascents have been made at Brussels, at Lyons, and at Amsterdam, under the direction of M. Nadar, assisted by a new lieutenant, M. Camille d'Artois, in the place of the Brothers Godard.

We have thus very rapidly sketched the life of a man whose name is known all over the world. Wonderfully endowed in every respect, tall and robust, active, lively, of an indomitable courage, and proverbially frank and honest, Nadar has scarcely the consciousness of his worth. Absolute and absorbing, he has touched upon almost every subject with success. As a man of letters, he has distinguished himself by the authorship of "Quand j'étais Etudiant," "La Robe de Déjanire," "Le Miroir aux Alouettes," "Les Mémoires du Géant," and, finally, of the "Droit au Vol," the French title of our "Right to Fly." As a caricaturist, he took the foremost place for wit and cleverness in the ranks of his brother artists; and now, faithful ever to the logic of his famous motto, "*Quand Même!*" and in spite of every discouragement and disappointment, he is busied in the pursuit of the study of that all-absorbing problem of the future, as to the ultimate successful solution of which M. Nadar is perfectly convinced.

PUSHKIN PRESS—THE LONDON LIBRARY

"FOUND ON THE SHELVES"

1   *Cycling: The Craze of the Hour*
2   *The Lure of the North*
3   *On Corpulence: Feeding the Body and Feeding the Mind*
4   *Life in a Bustle: Advice to Youth*
5   *The Gentlewoman's Book of Sports*
6   *On Reading, Writing and Living with Books*
7   *Through a Glass Lightly: Confession of a Reluctant Water Drinker*
8   *The Right to Fly*
9   *The Noble English Art of Self-Defence*
10 *Hints on Etiquette: A Shield Against the Vulgar*
11 *A Woman's Walks*
12 *A Full Account of the Dreadful Explosion of Wallsend Colliery by which 101 Human Beings Perished!*

THE LONDON LIBRARY (a registered charity) is one of the UK's leading literary institutions and a favourite haunt of authors, researchers and keen readers.

Membership is open to all.

Join at www.londonlibrary.co.uk.

www.pushkinpress.com